Student Activity Manual
for use with

Understanding Technology

by
R. Thomas Wright
Professor, Industry and Technology
Ball State University
Muncie, Indiana

and

Howard Bud Smith
Author/Chief Editor
Lee Howard Associates
Bayfield, Wisconsin

Publisher
The Goodheart-Willcox Company, Inc.
Tinley Park, Illinois

Introduction

This Student Activity Manual goes hand-in-hand with the **Understanding Technology** text.

Each Activity Manual chapter opens with an activity sheet. This sheet can be used with the corresponding text activity. However, the sheet is generic so you can use it to work out problems of design using your own imagination.

Most of the chapters have at least one alternate lab activity. Your instructor may assign these as extra credit or in place of text activities. Each chapter also contains a list of study questions corresponding to the text material.

Worksheets for text activities are also included in this manual. You will become familiar with writing up bills of material and procedures for activities.

Your response to questions allows your instructor to know where you might need additional help. You should study the chapter materials thoroughly before attempting to answer the study questions.

Safety should always be top priority in any activity. If you are not sure whether you are using materials and tools properly, check with your instructor. Always follow safety rules to avoid injury to yourself or others.

R. Thomas Wright
Howard Bud Smith

Copyright 1998

by

THE GOODHEART-WILLCOX COMPANY, INC.

Previous Edition Copyright 1993

International Standard Book Number 1-56637-375-1

2 3 4 5 6 7 8 9 10   98   01 00 99

Table of Contents

		MANUAL	TEXT
1	**What is Technology?**		7
	Activity (Procedure) Sheet	5	
	Laboratory Activity Follow-Up Sheet — Clay Tile Project	7	
	Alternate Laboratory Activity—Marble Counter	9	
	Study Questions	10	
2	**Tools and Technology**		24
	Activity Sheet—Safety	11	
	Alternate Laboratory Activity—Rubber Band Car	15	
	Alternate Laboratory Activity—Procedure and Tool Use	17	
	Study Questions	18	
3	**Materials and Technology**		47
	Activity Sheet—Applications	19	
	Laboratory Activity Data Recording Sheet—Material Properties	21	
	Alternate Activity—Product Analysis	22	
	Study Questions	23	
4	**Energy and Technology**		62
	Activity Sheet—Motor Schematic	25	
	Alternate Laboratory Activity—Wind-Powered Generator	27	
	Study Questions	29	
5	**Information and Technology**		88
	Activity Sheet—Note Taking	33	
	Alternate Laboratory Activity—Developing an Information Chart	35	
	Study Questions	36	
6	**People and Technology**		105
	Activity Sheet—Job Application	39	
	Laboratory Activity Sheet—Job Rating Form	41	
	Study Questions	43	
7	**Designing Technological Systems**		122
	Activity Sheet—Developing Design Ideas	45	
	Alternate Design Solution—Table Place Card	49	
	Study Questions	51	
8	**Manufacturing Systems**		141
	Activity Sheet—Process Demonstration	53	
	Alternate Laboratory Activity—Recipe Holder	55	
	Study Questions	57	

		MANUAL	TEXT
9	**Construction Technology**		**161**
	Activity Sheet—Sketching and Estimating	61	
	Alternate Design Activity—Model Bridge Contest	63	
	Study Questions	64	
10	**Communication Systems**		**193**
	Activity Sheet—Message Assessment	67	
	Laboratory Activity Sheet—Maze	69	
	Alternate Laboratory Activity—Billboard Message	70	
	Study Questions	71	
11	**Transportation Systems**		**219**
	Activity Sheet—Hull Design	75	
	Alternate Laboratory Activity — Rubber-Band Powered Boat	77	
	Study Questions	78	
12	**Production/Biorelated Systems**		**248**
	Activity Sheet — Bread Making Report	81	
	Alternate Laboratory Activity—Cheese Making	82	
	Study Questions	84	
13	**Impacts of Technology**		**256**
	Activity Sheet—Scrap/Waste Survey Chart	85	
	Alternate Laboratory Activity—Home Waste Report	87	
	Study Questions	88	
14	**Technology and the Future**		**270**
	Activity Sheet — Area Assignment	91	
	Alternate Activity—Future History	92	
	Alternate Laboratory Activity—Ocean Wave Power Generator	93	
	Study Questions	96	

What is Technology?

Activity (Procedure) Sheet

Name_____ Date _____

Period _____ Score _____

In the left column of this form, list the steps it will take to make the product as your teacher demonstrates this activity. As you complete the steps, record any observations or difficulties you have.

STEP — procedure	Describe difficulties in completing this step OR make observations about what you did or saw while you were completing the step.
1.	
2.	
3.	
4.	
5.	
6.	
7.	

Activity 2

Members of your group and their assignment:

Name	Assignment
1. _____	_____
2. _____	_____
3. _____	_____
4. _____	_____
5. _____	_____

On the chart below, list the steps you took to complete your part of the job.

1.
2.
3.
4.
5.
6.

How did the products produced in Activity 1 (no tooling) compare with those produced in Activity 2 (tooling)?

Laboratory Activity Follow-Up Sheet — Clay Tile Product

Name_____ Date _____

Period _____ Score _____

After completing activities for Chapter 1, pages 19 and 20, fill in the two follow-up Activity Sheets.

Activity 1 — Making a Tile Without Technology

PRODUCTIVITY:

A. Number of tiles produced by the class: _____

B. Number of workers (class members) _____

C. Productivity (output divided by workers) A/B _____

QUALITY: (Rate each element on a scale from 1 to 10 with 10 being the highest rating)

A. Uniformity: Thickness _____

               Shape _____

B. Surface Smoothness _____

C. Grooves: Position on the tile _____

           Size (width and depth) _____

Activity 2 — Making Tiles with Technology

PRODUCTIVITY:

A. Number of tiles produced by the class: _____

B. Number of workers (class members) _____

C. Productivity (output divided by workers) A/B _____

QUALITY: (Rate each element on a scale from 1 to 10 with 10 being the highest rating)

A. Uniformity: Thickness _____

               Shape _____

B. Surface Smoothness _____

C. Grooves: Position on the tile _____

           Size (width and depth) _____

On the back of this sheet, write a brief report comparing the results on these two activities.

COMPARISON: Activity 1 and Activity 2

_____

_____

_____

_____

_____

_____

_____

_____

_____

_____

_____

_____

_____

_____

_____

_____

_____

_____

_____

_____

_____

_____

_____

_____

_____

_____

_____

_____

_____

_____

_____

_____

_____

_____

chapter 1
What is Technology?

Alternate Laboratory Activity — Marble Counter

Name _____ Date _____

Period _____ Score _____

DESIGN BRIEF

You have a business that buys marbles from a manufacturer. It then places them in a bag for sale to toy stores. You need a device that will speed counting 10 marbles to be placed in a bag. On the grid below, sketch your solution to the problem.

-20-
marbles

Designed by: _____

CHAPTER 1 — Study Questions

Text pages: 7-21

Name _____

Date _____

Period _____ Score_____

1. Name five items you have which would not exist without technology:

   1. _____
   _____
   _____
   _____
   _____

2. _____A_____ is a study of nature and the laws that govern how the universe works. _____B_____ is the knowledge of how to control one's surroundings.

   2A. _____
   B. _____

MATCHING TEST. Match the definitions with parts of a technology system:

3. Resources used by a system.   A. Outputs.   3. _____
4. Reason for the system.   B. Processes.   4. _____
5. Actions taken to use the inputs.   C. Inputs.   5. _____
6. Changes made to processes to improve the outputs.   D. Goals.   6. _____
7. Result of the system.   E. Feedback.   7. _____
8. The technological processes that make up a system include (list all correct answers):   8. _____

   A. Energy processes.
   B. Production processes.
   C. Construction processes.
   D. Management processes.
   E. All of the above.

MATCHING TEST. Match the product or service with the technological system which produced it.

9. A trip to a distant city by air.   A. Construction.   9. _____
10. A new automobile.   B. Transportation.   10. _____
11. A problem solved by a computer.   C. Manufacturing.   11. _____
12. A new airport passenger terminal.   D. Communication.   12. _____

Tools and Technology

Activity Sheet — Safety

Name _____ Date _____

Period _____ Score _____

In the left column of this form, as your teacher demonstrates this activity, list the steps it will take to make the product. Record any safety procedures in the right column.

STEP	SAFETY PRECAUTION
1.	
2.	
3.	
4.	
5.	
6.	
7.	
8.	
9.	
10.	
11.	
12.	
13.	
14.	
15.	
16.	
17.	
18.	
19.	

STEP	SAFETY PRECAUTION
20.	
21.	
22.	
23.	
24.	
25.	
26.	
27.	
28.	
29.	
30.	
31.	
32.	
33.	
34.	
35.	
36.	
37.	
38.	
39.	
40.	
41.	
42.	

Record five steps using tools necessary to produce your product. List the tools. Mark each of the tools using the following codes:

M = measuring tool   Sl = slicing tool   H = holding tool
C = cutting tool   Sh = shearing tool   T = turning tool
Sa = sawing tool   D = drilling tool   Pd = pounding tool
                         G = gripping tool   Po = polishing tool

STEP	TOOLS USED	CODE

Tools and Technology

Alternate Laboratory Activity — Rubber Band Car

Name_____ Date _____

Period _____ Score _____

The following drawings show a product (rubber band powered vehicle). You may build it as you study about the common tools needed to make and use technological systems.

Working with your teacher, you should:

1. Develop a Bill of Materials (list) needed to make the product.

2. Prepare a procedure sheet (list of steps) for building the product. NOTE: List important safety rules with the procedure.

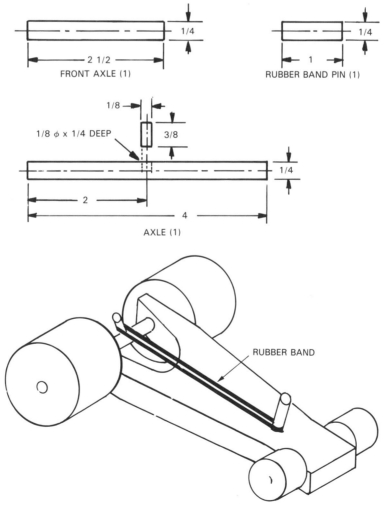

RUBBER BAND RACER

BODY

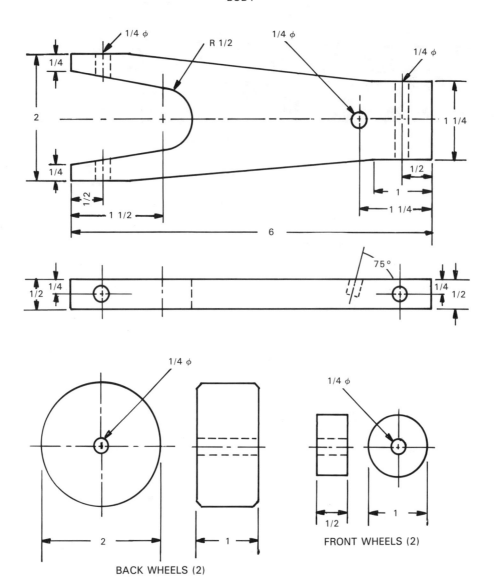

BACK WHEELS (2)

FRONT WHEELS (2)

chapter 2
Tools and Technology

Alternate Laboratory Activity — Procedure and Tool Use

Name_____ Date _____

Period _____ Score _____

The product you are building uses tools. They help you extend your ability to do work. On the chart below, select five steps you have completed. List the tools you used and indicate their type: (M = measuring; C = cutting; D = drilling; G = gripping; P = pounding and Po = Polishing).

STEP	TOOLS USED	TYPE
1.	a.	
	b.	
	c.	
	d.	
2.	a.	
	b.	
	c.	
	d.	
3.	a.	
	b.	
	c.	
	d.	
4.	a.	
	b.	
	c.	
	d.	
5.	a.	
	b.	
	c.	
	d.	

CHAPTER 2 — Study Questions

Text Pages: 24-46

Name _____

Date _____

Period _____ Score_____

1. What are primary tools and why are they necessary?

_____

_____

_____

MATCHING TEST. Match the descriptions with the proper list of tools.

2. Tools for communicating.   A. Scales, ruler, wristwatch.   2. _____

3. Tools used in business and commerce.   B. Hand truck, dolly, train, wheelbarrow.   3. _____

4. Tools used moving materials.   C. Calculators, computers, abacus.   4. _____

5. Tools used for measuring.   D. Computers, money, desks.   5. _____

6. Tools used for doing math.   E. Telephones, typewriters, computers, and pencils.   6. _____

7. The basic measuring unit in the SI metric system is the: 7. _____
   A. Decimal.
   B. Millimeter.
   C. Centimeter.
   D. Meter.
   E. Kilometer.

8. A scissors is an example of a _____-_____ lever.   8. _____

9. Properly label the parts of the first class lever shown below: 9A. _____
   B. _____
   C. _____

10. A person pedaling a bicycle is using the rear wheel as a: 10. _____
    A. Force multiplier.
    B. Distance multiplier.

Materials and Technology

Activity Sheet — Applications

Name _____ Date _____

Period _____ Score _____

You tested six specimens during this activity. Based on your test results describe each material and give three applications for the material.

DESCRIPTION	APPLICATIONS
Specimen #1	1. 2. 3.
Specimen #2	1. 2. 3.
Specimen #3	1. 2. 3.
Specimen #4	1. 2. 3.
Specimen #5	1. 2. 3.
Specimen #6	1. 2. 3.

Materials and Technology

Laboratory Activity Data Recording Sheet — Material Properties

Group Number: _____

Members: (a) _____ (b) _____

(c) _____ (d) _____

After your group has tested the materials given you for your Chapter 3 Activity (pages 59-61), describe their properties on the chart below. Write a brief statement which describes the properties listed for each specimen.

Specimen Number	1	2	3	4	5	6
Name of Material						
Type of Material						
Density						
Hardness						
Light Reflectivity						
Torsion Strength						

chapter 3
Materials and Technology

Alternate Laboratory Activity — Product Analysis

Name_____ Date _____

Period _____ Score _____

Assignment
1. Select any product in the technology education laboratory. 2. List four materials that were used to make the product. 3. On the chart below, list the material and its use. 4. Indicate on the chart why you think the product designer chose each material used.

Product chosen: _____

1. Material used:

   _____

   Where it was used:

   _____

   Why you think it was chosen:

2. Material used:

   _____

   Where it was used:

   _____

   Why you think it was chosen:

3. Material used:

   _____

   Where it was used:

   _____

   Why you think it was chosen:

4. Material used:

   _____

   Where it was used:

   _____

   Why you think it was chosen:

CHAPTER 3 — Study Questions

Text Pages: 47-61

Name _____

Date _____

Period _____ Score_____

1. Technological systems use:
   A. Two types of materials.
   B. Three types of materials.
   C. Four types of materials.
   D. No materials.

   1. _____

2. Products, such as a chair, have a set form or shape.

   Therefore they must be made from _____ _____.

   2. _____

3. What do metals, ceramics, polymers, and composites have in common?

   _____

4. A basic name for materials that have never been alive is:
   A. Ceramics.
   B. Metals.
   C. Inorganic materials.
   D. Engineering materials.

   4. _____

5. The photograph below shows:
   A. Organisms in living material.
   B. Magnified grain structure of metal.
   C. Ceramic crystals magnified.

   5. _____

6. Glass belongs to a class of engineering materials known as:
   A. Composites.
   B. Crystals.
   C. Ceramics.
   D. Plastics.

   6. _____

7. Rubber and plastics belong to the same general classs of materials known as polymers. True or false?

7. _____

8. Indicate whether the following materials are exhaustible or renewable.

   A. Wood.

   B. Rubber.

   C. Animals.

   D. Corn.

   E. Iron.

   F. Coal.

   G. Soil.

7A. _____

B. _____

C. _____

D. _____

E. _____

F. _____

G. _____

9. Name two methods of extraction used to get exhaustible materials from the earth.

9. _____
_____

10. How a material reacts to forces or conditions is called its _____ properties.

10. _____

Energy and Technology

Activity Sheet — Motor Schematic

Name_____ Date _____

Period _____ Score _____

Draw a schematic for your motor on the grid below.

List below three ways you could improve the efficiency of your motor:

(Use reverse of page if additional space is needed.)

chapter 4
Energy and Technology

Alternate Laboratory Activity — Wind-Powered Generator

Name_____ Date _____

Period _____ Score _____

Given a small, low voltage, direct current motor, design a working wind-powered electric generating system. Your design will be compared with other designs produced in your class to determine the most efficient system. Use the reverse side for your final sketch.

TWO EXAMPLES FOR DESIGNS

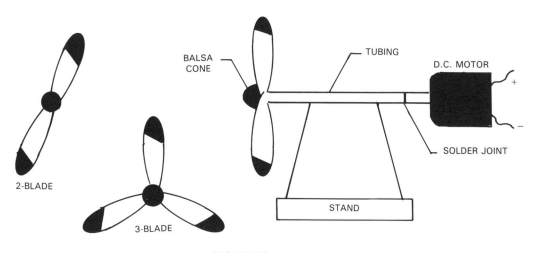

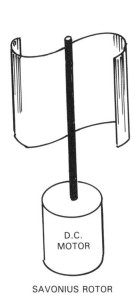

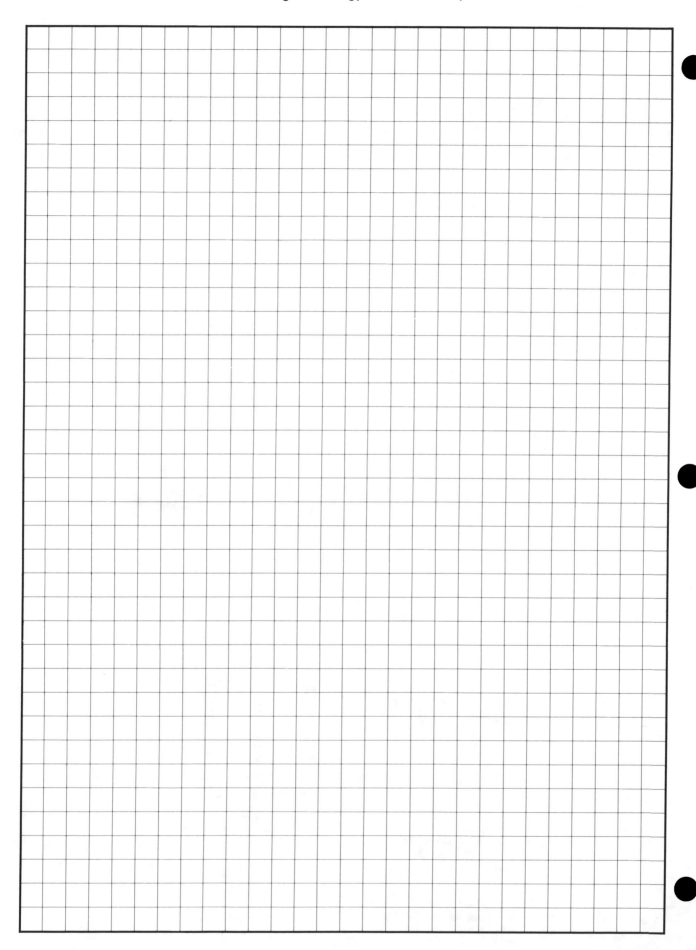

CHAPTER 4 — Study Questions

Text Pages: 62-87

Name _____

Date _____

Period _____ Score_____

1. No industry is possible without energy. True or false?

   1. _____

2. Energy is the _____ to do _____.

   2. _____

   _____

3. Power and energy are (select all correct answers):
   A. The same thing.
   B. Energy is ability to do work while power is force times distance divided by time.
   C. Energy is ability to do work while power is the speed at which work is done.

   3. _____

4. Energy in motion is called _____ energy.

   4. _____

5. List and describe six forms of energy.

_____

_____

_____

_____

_____

_____

_____

_____

_____

_____

_____

_____

6. It is the carbon in fossil fuels that makes them burn so easily. True or false?

   6. _____

7. Why is water power considered a renewable source of energy?

_____

_____

8. An electrical generator is classified as an ___A___ ___B___ because it uses one kind of energy and changes it to another type.

   8A. _____

   B. _____

9. Look at the illustration and tell what process it represents.

9. _____

_____

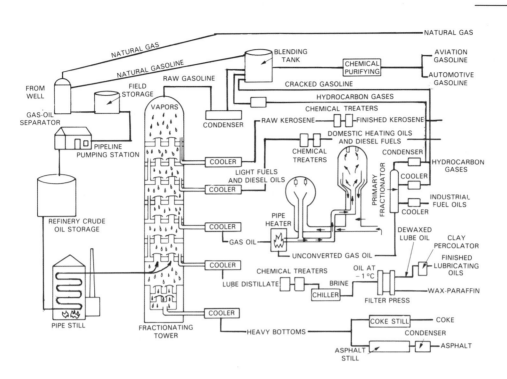

10. Look at the illustration and tell what is happening.

10. _____

_____

_____

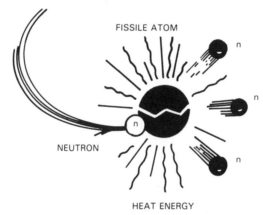

11. The illustration represents a type of energy converter. Can you name it?

11. _____

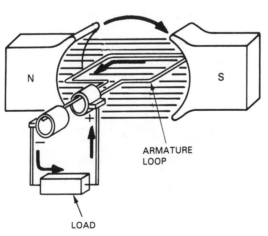

12. What type of energy does the device in Question 11 produce?   12. _____

13. Look at the illustration and tell what it is.   13. _____

14. Tell what part of its cycle the above drawing is entering.   14. _____
_____

15. A ___A___ ___B___ converts chemical energy directly into   15A. _____
electrical energy.   B. _____

16. Moving power from one place to another is called _____.   16. _____

17. List six ways that power can be transmitted.

_____
_____
_____

18. Putting in thicker insulation on building walls and ceilings   18. _____
will save energy. True or false?

19. Summarize what you have learned about energy in Chapter 4 of the text. (Use space below and on the back of this sheet.)

_____
_____
_____
_____
_____
_____
_____
_____

Information and Technology

Activity Sheet — Note Taking

Name_____ Date _____

Period _____ Score _____

Take notes from the brief articles on pages 102-104 below.

(Over)

Prepare an outline for your report below.

Prepare a sketch of any illustrations or charts you will use in your report below.

Information and Technology

Alternate Laboratory Activity — Developing an Information Chart

Name _____ Date _____

Period _____ Score _____

The "Applying Your Knowledge" activity on pages 43-45 of the text provides you with information you need to make the product. On the chart, list the types of information you would need to build the product and where that information was given (D = drawings, B = bill of materials, P = procedures).

INFORMATION NEEDED	WHERE AVAILABLE	SAMPLE OF THE INFORMATION (copy of typical entry)
1.		
2.		
3.		
4.		

CHAPTER 5 — Study Questions

Text Pages: 88-104

Name _____

Date _____

Period _____ Score_____

1. What is the difference between knowledge and theory?

_____

_____

_____

2. List the three ideas that are the basis for the Scientific Method.

_____

_____

_____

_____

_____

_____

_____

_____

_____

_____

3. Measurement is not important to scientific research.   3. _____
   True or false?

4. _____ sciences deal with matter in its purest form.   4. _____

5. _____ sciences are concerned with history, properties and   5. _____
   composition of material, and behavior of the natural world.

6. A study of living matter is called: 6. _____
   A. Geology.
   B. Oceanography.
   C. Life science.
   D. Physical science.

7. Explain the meaning of "appropriate technology."

_____

_____

8. Throughout history, technology has benefitted people. Briefly list three ways in which it has helped.

_____

_____

_____

9. _____ deals with human culture and backgrounds.   9. _____

10. The following are essential parts of a computer system. Explain what each part does:

A. Input device. _____

_____

B. Processor. _____

_____

C. Memory. _____

_____

D. Output device. _____

_____

11. Basically, a computer makes decisions by giving either a "yes"   11. _____
or a "no" answer. True or false?

12. ROM is a computer term which means (pick best answer):   12. _____
A. Ready on Monday.
B. Read only memory.
C. Is not a computer term.
D. Contains all the data fed into the computer.

13. List all the sources of information that you use in a typical week. Then write a report on how you find the information useful. Consider that some of the information is immediately useful (something you can act on) while other information may be stored for future use. (Use reverse side of this page, if necessary.)

_____

_____

_____

_____

_____

_____

_____

_____

_____

_____

_____

_____

_____

_____

_____

Activity Sheet — Job Application

Name _____ Date _____

Period _____ Score _____

Fill out the application for a job. On the reverse side, answer the questions.

APPLICATION FOR EMPLOYMENT

To the applicant: We appreciate your interest in our organization and are interested in your qualifications. An understanding of your background and work history will aid us in placing you in a position which best meets your qualifications.

PERSONAL

Name: _____ Social Security No. _____
(Last)   (First)   (Middle Initial)

Address: _____ Telephone Number: _____

How long have you lived at the above address? _____

EMPLOYMENT RECORD

Position Desired: _____ Date Available: _____

List all past employers with the most recent first. (Include summer and part time positions.)

EMPLOYER	DATES EMPLOYED FROM TO		JOB TITLE	REASON FOR LEAVING

EDUCATION RECORD

EDUCATIONAL INSTITUTION	DATES FROM TO		COMPLETED YES NO	
Grade School				
Junior High School				
High School				

REFERENCES

NAME	ADDRESS	OCCUPATION

After you have filled out the Application for Employment, answer the following questions drawing on your own understanding of the purpose of the application. (Try to place yourself in the role of the employer before you answer the questions.)

Why does the employer ask for personal information?

_____

_____

_____

Why do you suppose a prospective employer would ask for a list of your previous employers and reasons for leaving?

_____

_____

_____

_____

_____

_____

Suggest a reason or reasons why an employer would want a record of your education.

_____

_____

_____

_____

_____

Why would an employer ask for references?

_____

_____

_____

If you were hiring someone, would you want any additional information about the applicant? If so, list the information desired.

_____

_____

_____

People and Technology

Laboratory Activity Sheet — Job Rating Form

Name_____ Date _____

Period _____ Score _____

(Refer to Activity on page 117 of your text before filling in the form.)

JOB RATING FORM

Job Title: _____

People Skills:	_____	Mental Skills:	_____
Physical Skills:	_____	Management Skills:	_____
Artistic Skills:	_____	Communication Skills:	_____

Job Title: _____

People Skills:	_____	Mental Skills:	_____
Physical Skills:	_____	Management Skills:	_____
Artistic Skills:	_____	Communication Skills:	_____

Job Title: _____

People Skills:	_____	Mental Skills:	_____
Physical Skills:	_____	Management Skills:	_____
Artistic Skills:	_____	Communication Skills:	_____

Job Title: _____

People Skills:	_____	Mental Skills:	_____
Physical Skills:	_____	Management Skills:	_____
Artistic Skills:	_____	Communication Skills:	_____

Job Title: _____

People Skills:	_____	Mental Skills:	_____
Physical Skills:	_____	Management Skills:	_____
Artistic Skills:	_____	Communication Skills:	_____

JOB RATING FORM (continued)

Job Title: _____

People Skills:	_____	Mental Skills:	_____
Physical Skills:	_____	Management Skills:	_____
Artistic Skills:	_____	Communication Skills:	_____

Job Title: _____

People Skills:	_____	Mental Skills:	_____
Physical Skills:	_____	Management Skills:	_____
Artistic Skills:	_____	Communication Skills:	_____

Job Title: _____

People Skills:	_____	Mental Skills:	_____
Physical Skills:	_____	Management Skills:	_____
Artistic Skills:	_____	Communication Skills:	_____

Job Title: _____

People Skills:	_____	Mental Skills:	_____
Physical Skills:	_____	Management Skills:	_____
Artistic Skills:	_____	Communication Skills:	_____

Job Title: _____

People Skills:	_____	Mental Skills:	_____
Physical Skills:	_____	Management Skills:	_____
Artistic Skills:	_____	Communication Skills:	_____

Job Title: _____

People Skills:	_____	Mental Skills:	_____
Physical Skills:	_____	Management Skills:	_____
Artistic Skills:	_____	Communication Skills:	_____

Job Title: _____

People Skills:	_____	Mental Skills:	_____
Physical Skills:	_____	Management Skills:	_____
Artistic Skills:	_____	Communication Skills:	_____

CHAPTER 6 — Study Questions

Text Pages: 105-119

Name _____

Date _____

Period _____ Score_____

1. Name five types of communication that make it easier for you   1. _____
   to get news and information.

   _____

   _____

   _____

   _____

   _____

2. What do physical characteristics have to do with the kind of work you may choose?

   _____

   _____

3. Name three places likely to have literature or books on careers.   3. _____

   _____

   _____

4. Name one type of job or career you think you would like and tell why: _____

   _____

   _____

   _____

   _____

   _____

   _____

5. Give two meanings for the word "enterprise."

   _____

   _____

6. Name one trait desirable in a manager.

   _____

   _____

   _____

   _____

7. Why is it said that persons entering the job market must be flexible?

   _____

   _____

   _____

8. List three responsibilities that are assumed by an entrepreneur:
   A. _____
   B. _____
   C. _____

9. Most people become entrepreneurs because it usually means they work fewer hours. True or false?

   9. _____

10. Assume that you are forming a new company. In the space below draw an organizational chart for company officers. Fill in the names of the management positions. (Refer to Fig. 6-8 in the text.) Note: You may have fewer or more management positions than are shown in Fig. 6-8.

_____
_____
_____
_____
_____
_____
_____
_____
_____
_____

Activity Sheet — Developing Design Ideas

Name _____ Date _____

Period _____ Score _____

Develop a set of criteria for your greeting card below:

Event: _____

Audience: _____

Develop three to five messages that could be used for the card below:

_____

_____

_____

_____

_____

Develop rough sketches for the outside of three to five cards on the grids below and on the next page:

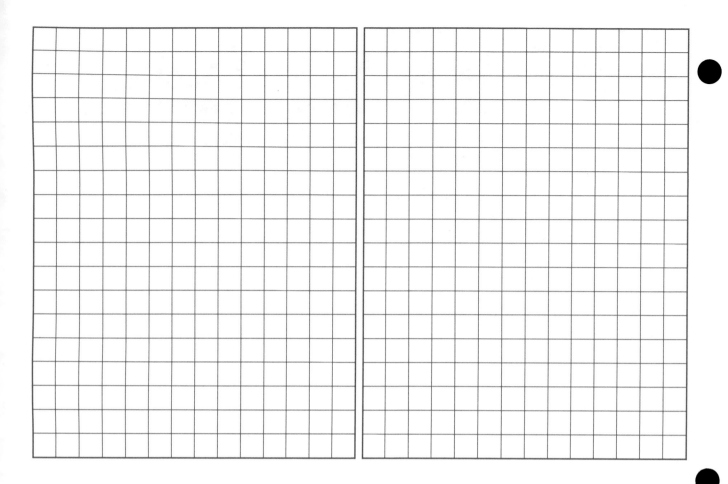

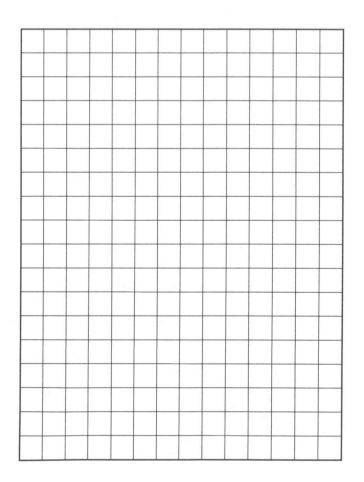

Make a final layout of your greeting card below.

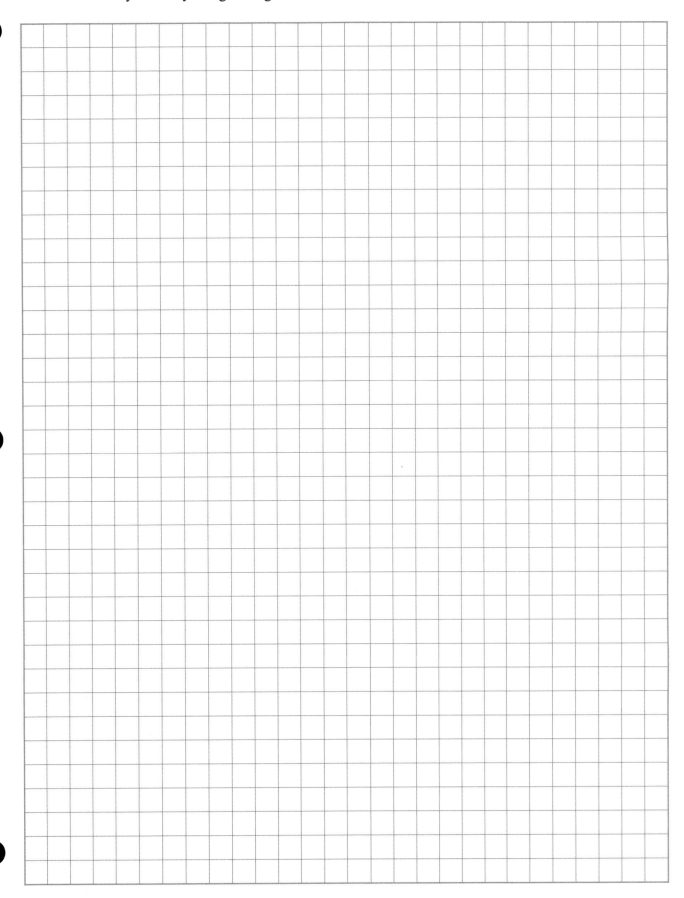

Often, before any design work is attempted, the designer collect information from potential buyers of the product. In the space below develop some questions for a questionnaire you would use to determine what would be a popular design for a birthday card.

_____

_____

_____

_____

_____

_____

_____

_____

_____

_____

_____

_____

_____

_____

_____

_____

_____

_____

Designing Technological Systems

Alternate Design Solution — Table Place Card

Name _____ Date _____

Period _____ Score _____

DESIGN BRIEF

You have been employed to design a table place card. This card will be placed on the table at a school recognition banquet so people will know where to sit. Select an event, such as a music awards banquet, or a football team recognition banquet. Then design the card in the space below. The card should have a cut-out so the symbol for the event will stand out. The example on the left shows a design for a baseball awards banquet.

ROUGH SKETCHES

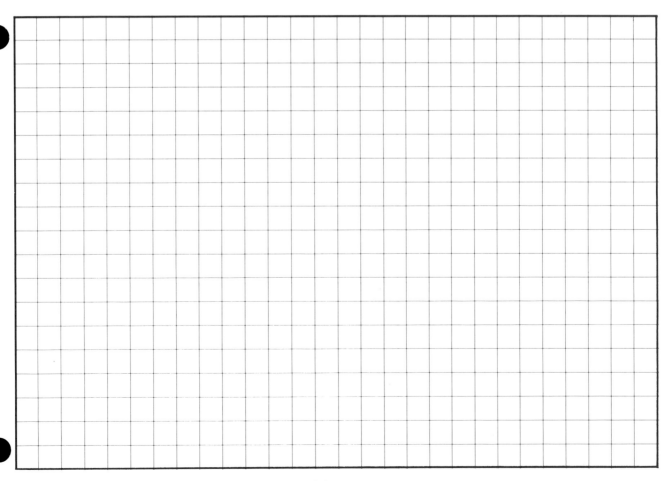

Designed by: _____

REFINED SKETCHES

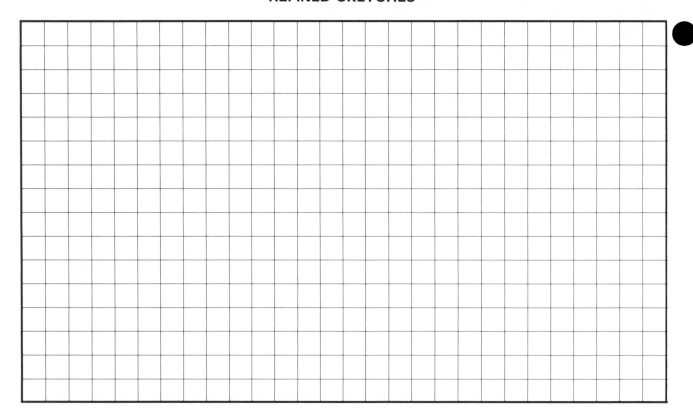

DETAILED SKETCHES

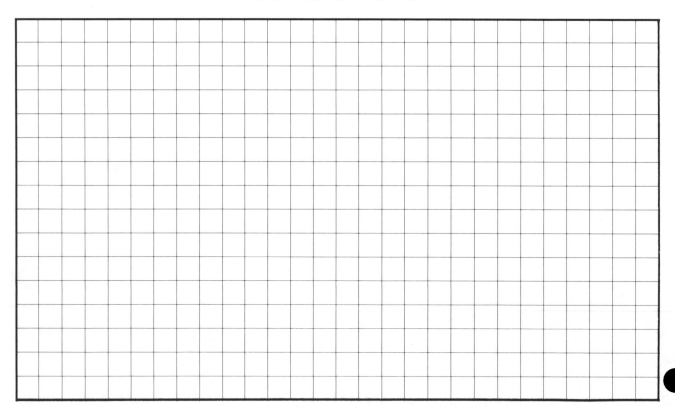

Drawn by: _____

CHAPTER 7 — Study Questions

Text Pages: 122-140

Name _____

Date _____

Period _____ Score_____

1. Look at the illustration and fill in the missing design steps.

1A. _____

B. _____

C. _____

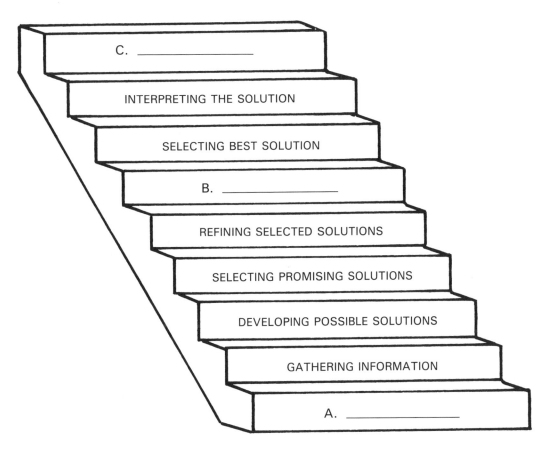

C. _____

INTERPRETING THE SOLUTION

SELECTING BEST SOLUTION

B. _____

REFINING SELECTED SOLUTIONS

SELECTING PROMISING SOLUTIONS

DEVELOPING POSSIBLE SOLUTIONS

GATHERING INFORMATION

A. _____

2. When designers have trouble developing good solutions, they may need to collect more information from research. True or false?

2. _____

3. Using information from a later design step to improve an earlier step is called:
   A. Brainstorming.
   B. Testing design solutions.
   C. Redesign.
   D. Feedback.

3. _____

4. In design, the design _____ gives all the other design steps proper direction.

4. _____

5. Look at the chart below and determine to what designing activity the actions belong.

5A. _____

_____

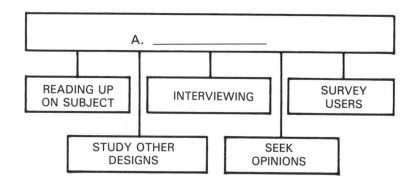

6. When you ask many people's opinion about a proposed new product, you are doing ___A___ ___B___.

6A. _____

B. _____

7. The drawing shown is:
   A. A refined sketch.
   B. Doodling.
   C. A detailed sketch.
   D. A rough sketch.

7. _____

_____

8. In a (an) _____ drawing, you first draw a vertical line. Then you draw other lines at a 30-degree angle to the vertical line.

8. _____

9. Mock-ups are working models of a new product. True or false?

9. _____

10. A complex ___A___ ___B___ can test a product without having an actual model.

10A. _____

B. _____

11. The ___A___ of a new product is to meet the stated ___B___; that is, the reason for its design.

11A. _____

B. _____

12. The bill of materials is a major design communication device. True or false?

12. _____

Manufacturing Systems

Activity Sheet — Process Demonstration

Name_____ Date _____

Period _____ Score _____

During the demonstration of the processes to be used to manufacture your product, complete all three columns of this chart. In column 1 write a very brief description of each step. In column 2 list the tools or machines that will be used. In column 3 list any safety precautions you should observe.

STEP	TOOL OR MACHINE	SAFETY CONSIDERATION
1.		
2.		
3.		
4.		
5.		
6.		
7.		
8.		
9.		
10.		
11.		
12.		

STEP	TOOL OR MACHINE	SAFETY CONSIDERATION
13.		
14.		
15.		
16.		
17.		
18.		
19.		
20.		
21.		
22.		
23.		
24.		
25.		
26.		
27.		
28.		
29.		

Alternate Laboratory Activity — Recipe Holder

Name_____   Date _____

Period _____   Score _____

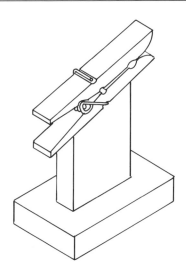

ASSIGNMENT

Study the drawings for the recipe holder. Then draw up a Bill of Materials for it. On the following page, list the procedure for making the base and the procedures for making the upright.

BILL OF MATERIALS				
QTY.	PART NAME	SIZE		
		Thickness	Width	Length

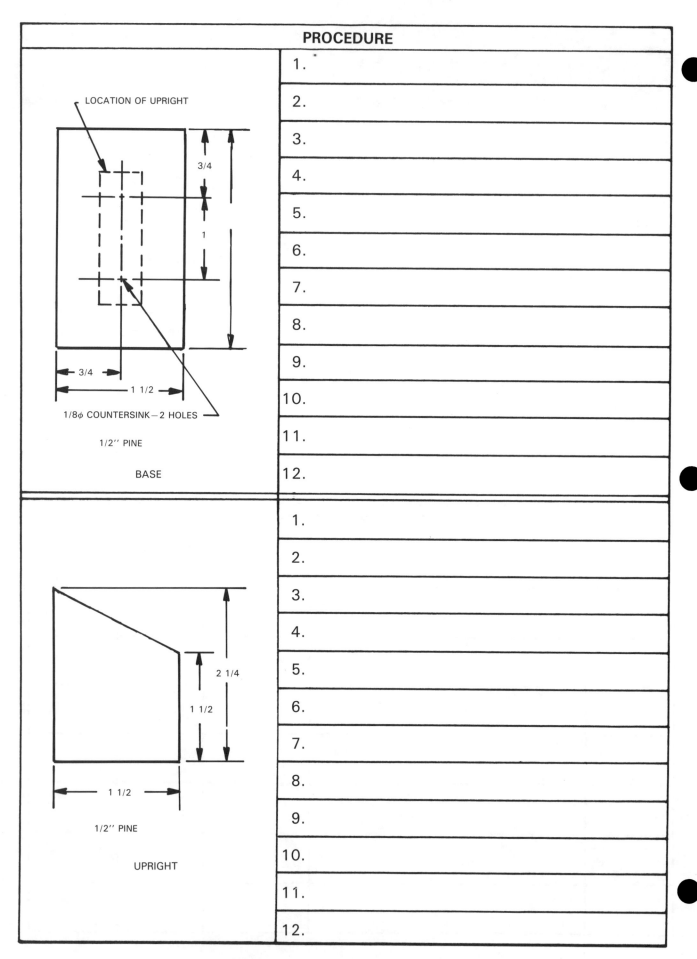

PROCEDURE

1.
2.
3.
4.
5.
6.
7.
8.
9.
10.
11.
12.

LOCATION OF UPRIGHT

3/4

1

3/4

1 1/2

1/8φ COUNTERSINK—2 HOLES

1/2'' PINE

BASE

1.
2.
3.
4.
5.
6.
7.
8.
9.
10.
11.
12.

2 1/4

1 1/2

1 1/2

1/2'' PINE

UPRIGHT

CHAPTER 8 — Study Questions

Text Pages: 141-160

Name _____

Date _____

Period _____ Score_____

1. Fill in the blanks in the illustration below:

1A. _____

   B. _____

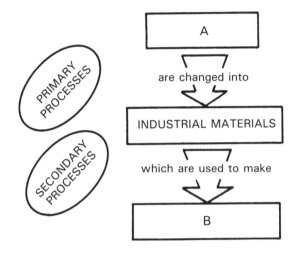

2. Changing lumber into skateboards is an example of primary processing. True or false?

2. _____

3. The output of primary processing is called ___A___ ___B___.

3A. _____

   B. _____

4. Identify the parts of the process shown below:

4A. _____

   B. _____

   C. _____

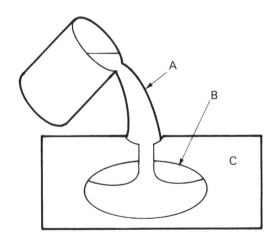

5. When a new mold is used for each product, the mold is called a (an) ___A___ ___B___ mold.

5A. _____

   B. _____

6. In thermoforming, _____ helps the material take its proper shape.

6. _____

7. Name the process shown below:

7. _____

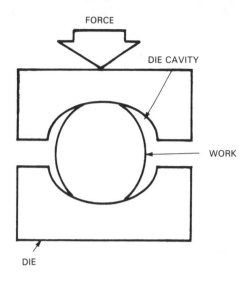

8. Fill in the blanks for the cutting operation shown:

8A. _____
   B. _____
   C. _____

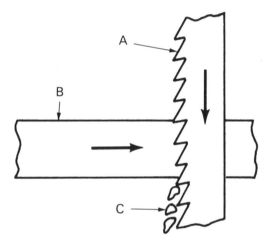

9. In _____ cutting, burning gases melt away unwanted material.

9. _____

10. Look at the illustration and name the process.

10. _____

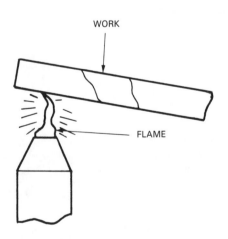

11. The illustration shows a (an) ___A___ ___B___.

11A. _____

B. _____

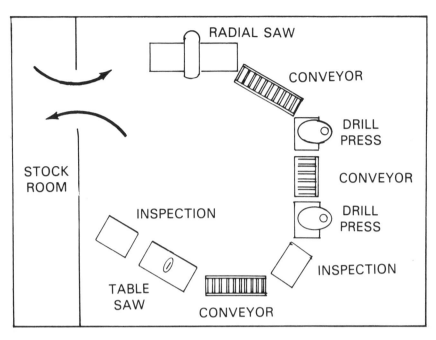

12. Making a production run to see that the manufacturing system is working correctly is called:
   A. Intermittent manufacturing.
   B. Custom manufacturing.
   C. A pilot run.
   D. A test run.

12. _____

13. Name four ways advertising reaches people who might buy a new product.

13. _____

_____

_____

_____

14. What three factors might cause a customer to buy a product?

_____

_____

_____

15. List as many manufactured items as you can that you find useful. Tell how they benefit you and make your life easier and/or more pleasing. If you can suggest what process was used to produce each. (Consider: casting, forming, separating, conditioning, finishing and assembling. Remember, many products may require several processes.) Use reverse side of this page if more room is needed.

_____

_____

_____

_____

_____

_____

_____

_____

_____

_____

_____

_____

_____

_____

_____

_____

_____

_____

_____

_____

Construction and Technology

Activity Sheet — Sketching and Estimating

Name_____ Date _____

Period _____ Score _____

Name of the Construction Project: _____

Part of the structure your group is producing: _____

Date to be completed: _____

On the grid below, draw a sketch of the building component you are building. Label all the major parts.

ESTIMATING CONSTRUCTION COSTS

Estimating the cost of a structure requires making educated guesses about how much material will be needed, how much labor is involved, and what the land will cost. **Labor costs** are estimated by multiplying the number of workers by the number of hours required to complete the structure. **Equipment costs** are based on money that must be spent to rent or buy tools, machines, and other equipment. **Material costs** are arrived at by checking catalog prices or checking with a lumber yard. **Land costs** can be determined by checking real estate ads in local newspapers.

Your assignment is to determine the cost of building the shed in your textbook activity. Place your estimates in the chart. Use the equipment and supply list on page 189 of your text.

CONSTRUCTION ESTIMATE FOR STORAGE SHED		
ITEM		**ESTIMATE**
Land		
Labor (five crew members)		
Equipment (list all tools on page 189)		
QUANTITY	**DESCRIPTION**	
Materials		
QUANTITY	**DESCRIPTION**	

Construction and Technology

Alternate Design Activity — Model Bridge Contest

Name _____ Date _____

Period _____ Score _____

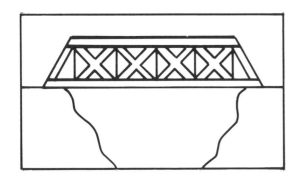

DESIGN BRIEF

Your task is to design a 12 in. long x 2 in. high model bridge. You will have 10 pieces of 1/4" x 1/4" x 12" material. You may use any design but your model will be compared with others built in the class. The bridge which holds the most weight is the best design. Each bridge will be tested to its breaking strength on a test stand similar to the one at the bottom of this sheet.

PROCEDURE

1. Research bridge designs in the school library.
2. Design and draw a full-scale bridge framework.
3. Cut the parts needed to make the two sides of the bridge.
4. Glue the bridge sides together.
5. Assemble the two sides with spacers that hold the sides 1 1/2 in. apart.
6. Test the strength of the bridge, recording the maximum weight it supported.
7. Compare the results of your bridge with those of other class members.
8. Determine the "best design" (the bridge holding the heaviest load before breaking).

TEST STAND

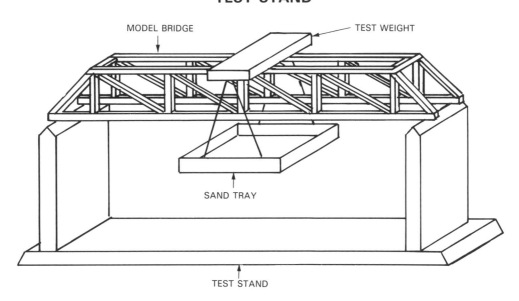

MODEL BRIDGE   TEST WEIGHT

SAND TRAY

TEST STAND

CHAPTER 9 — Study Questions

Text Pages: 161-192

Name _____

Date _____

Period _____ Score_____

1. _____ organizes construction resources and uses them efficiently.

    1. _____

2. The most famous builders among ancient people were the _____, _____, and _____.

    2A. _____

    B. _____

    C. _____

3. Civil structures are houses built for a single family's use. True or false?

    3. _____

4. Pipes and cables that supply buildings with water, fuel, and electricity are generally known as _____.

    4. _____

MATCHING TEST. Match the term with the right description:

5. Designs the structure.

6. Hires and directs skilled workers.

7. Adds up probable construction costs.

8. Helps contractor. Sees that construction work is done properly.

A. Estimator.

B. Project manager.

C. Architect.

D. Contractor.

    5. _____

    6. _____

    7. _____

    8. _____

9. The drawing shown below is called a (an) _____ _____.

    9. _____

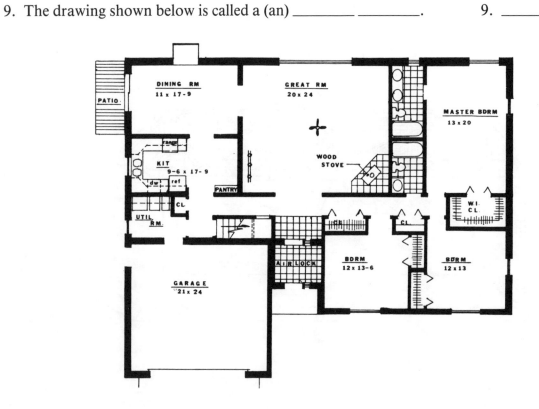

10. This drawing is called a (an) _____.   10. _____

11. _____ is a procedure a government uses to acquire land that the owner does not want to sell.   11. _____

12. Soil must be tested for its ability to support a structure. In a _____ test, you would place the soil in a cylinder. Then you'd compact it with a hammer.   12. _____

13. Making sure that a structure is strongly built is not the designer's job. True or false?   13. _____

14. Sometimes a contractor will pass along part of the work of construction to other companies. These companies are known as _____.   14. _____

15. Fill in the blanks in the construction management chart below.   15.A. _____

B. _____

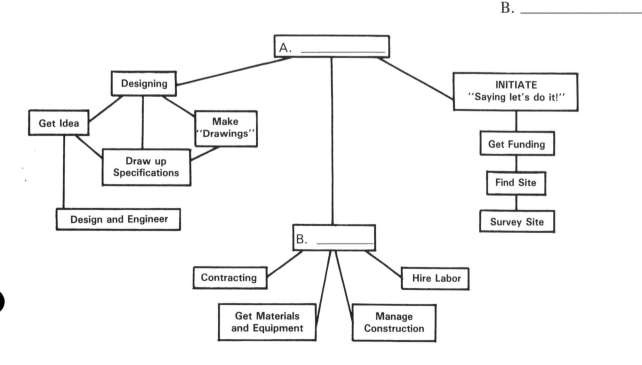

16. Before any construction work is done the owner or contractor for a new building must get a (an):
    A. Occupancy permit.
    B. Soil test.
    C. Building permit.
    D. Builder's license.

16. _____

17. Look at the three types of foundations and name each type:

17A. _____

B. _____

C. _____

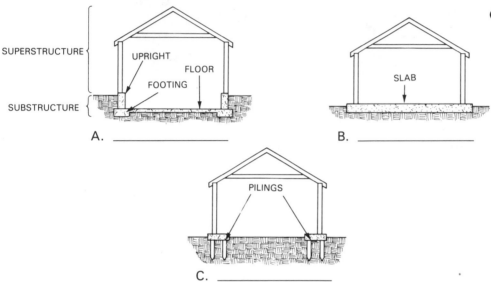

A. _____

B. _____

C. _____

18. Look at the illustration below and tell what part of the building it is.

18. _____

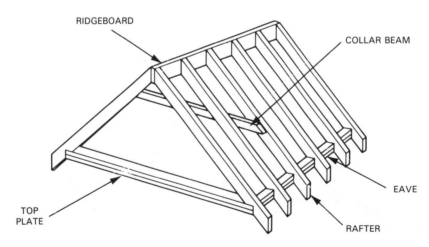

19. Installing mechanical systems in a building is called _____.

19. _____

20. When you paint and decorate you are doing the _____ work on a structure.

20. _____

21. _____ is a way of making a site more attractive. It includes planting trees, shrubs, and flowers.

21. _____

Communication Systems

Activity Sheet — Message Assessment

Name_____ Date _____

Period _____ Score _____

Complete Assignment 1 from page 216 of the text and then answer the questions below:

Was your group able to decode the message sent to you? _____

What was the message? _____

_____

_____

Describe what you think is the input, process, and output of the telegraph system. _____

_____

_____

_____

What was the distance over which the message was sent? _____

How could the telegraph made by your group be modified so that a deaf person could use it?

_____

_____

_____

Working as a group, modify the design to give a visual message.

Describe what was done and tell if it worked. _____

_____

_____

_____

_____

_____

(Place a sketch of your modification on the reverse side of this page.)

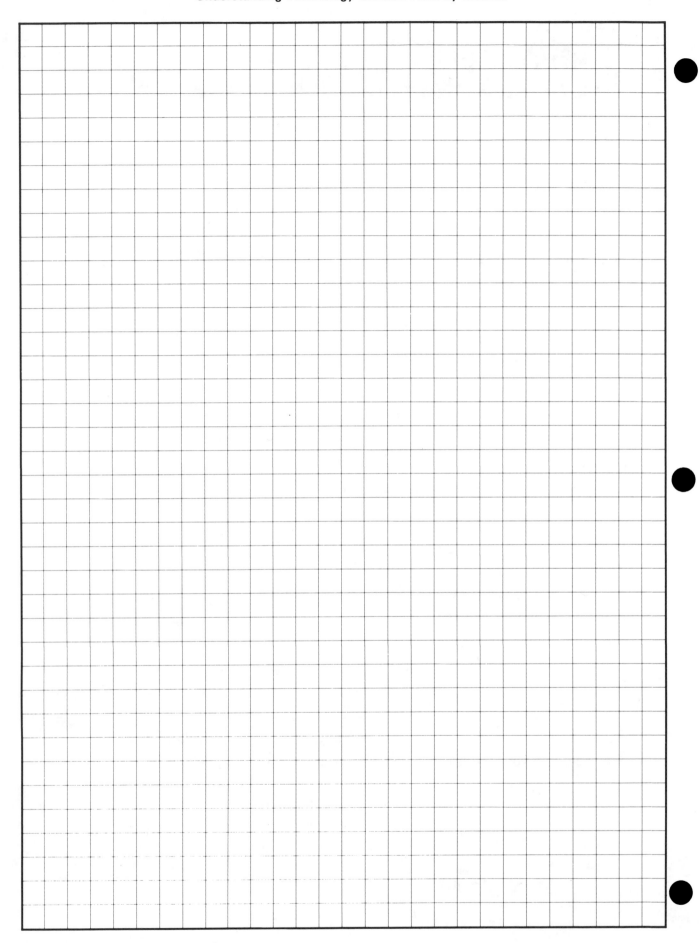

chapter 10
Communication Systems

Laboratory Activity Sheet — Maze

Name _____ Date _____

Period _____ Score _____

INSTRUCTIONS: Read the maze activity on pages 215-218 of your text. Then solve the maze below. ("A" groups will solve the "A" maze. "B" groups will solve the "B" maze.

Group A Maze

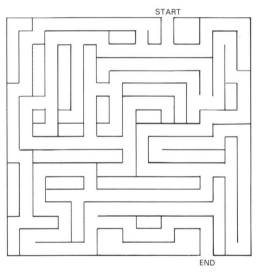

Group B Maze

chapter 10
Communication Systems

Alternate Laboratory Activity — Billboard Message

Name _____ Date _____

Period _____ Score _____

DESIGN BRIEF

Graphic designers must be concerned about using printed messages to attract readers. They must determine who their audience is and how to get their attention. They must then design an effective message. Select one of the design problems below and develop it for the audience specified.

Problem No. 1:

Design a billboard message that will promote safe bicycle riding habits among students in your schools.

Problem No. 2:

Recently a problem of trash on the school grounds has developed. Design a poster that will address this problem. Direct the message to the students in your school.

MESSAGE DESIGN

Theme or slogan to be used: _____

Message layout:

CHAPTER 10 — Study Questions

Text Pages: 193-218

Name _____

Date _____

Period _____ Score_____

1. Write a short report (below) on what one day at school would be like if no one communicated.

_____
_____
_____
_____
_____
_____
_____
_____
_____
_____
_____
_____
_____
_____
_____
_____

2. What is the difference between communication and communication technology?

_____
_____
_____

3. The two basic types of communication systems are:
   A. Broadcast systems and printing systems.
   B. Graphic systems and wave systems.
   C. Electronic systems and photographic systems.

3. _____

4. Use of drawings, pictures, graphs, photographs, or words printed on flat surfaces is known as _____ communications.

4. _____

5. All electromagnetic waves travel at the speed of:
   A. Sound—1100 ft. per second at sea level.
   B. Light—186,000 miles per second.
   C. Light—186,000 feet per second.
   D. Electricity—which cannot be measured.

5. _____

6. The illustration below shows the electromagnetic spectrum. Identify the indicated waves.

6A. _____

B. _____

C. _____

D. _____

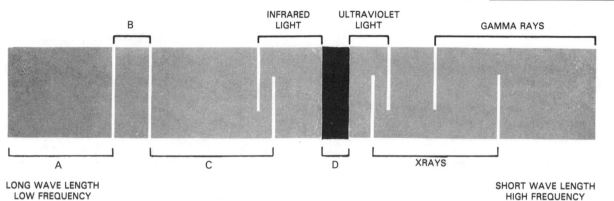

B

INFRARED LIGHT

ULTRAVIOLET LIGHT

GAMMA RAYS

A

C

D

XRAYS

LONG WAVE LENGTH
LOW FREQUENCY

SHORT WAVE LENGTH
HIGH FREQUENCY

7. Through fiber optics, sound can be changed into pulses of _____.

7. _____

8. What is the chief advantage of electronic communication over graphic communication?

_____

_____

_____

9. Name the parts of the simple communication system shown below:

9A. _____

B. _____

C. _____

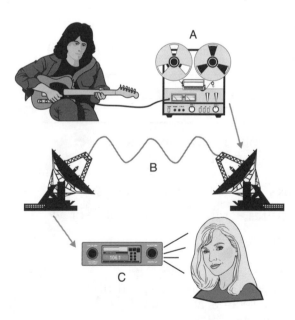

A

B

C

Communication Systems

10. Name the types of communication pictured below.

10A. _____
B. _____
C. _____
D. _____

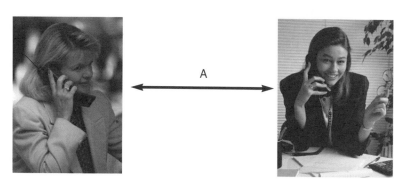

A

B

C

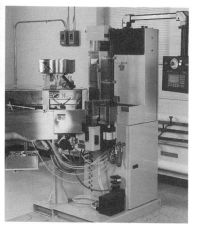

D

11. If an orbiting communication satellite always stays above the same spot on earth, its orbit is called _____.

11. _____

12. Look at the illustration below. Name the parts.

12A. _____
B. _____
C. _____

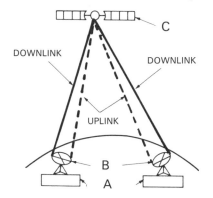

13. Name the printing process shown below.

13. _____

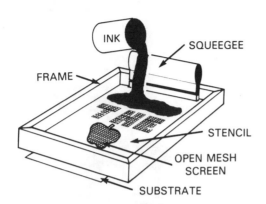

14. List three ways in which communication can be stored.

_____

_____

_____

15. _____ is a printing process which uses an electrically charged drum.

15. _____

16. The written record of the words and actions for a film message is called a _____.

16. _____

Transportation Systems

Activity Sheet — Hull Design

Name _____ Date _____

Period _____ Score _____

Type of hull: _____

On the grid below, draw a sketch of the hull you will build for your boat. Dimension the major distances.

HULL TEST DATA

Record the data from the tests of the three hull shapes. Rank them in terms of their efficiency.

HULL DESIGN #1

Type of design: _____

Sketch shape below

Travel time:

Test #1 _____

Test #2 _____ RANK

Test #3 _____

Average _____

HULL DESIGN #2

Type of design: _____

Sketch shape below

Travel time:

Test #1 _____

Test #2 _____ RANK

Test #3 _____

Average _____

HULL DESIGN #3

Type of design: _____

Sketch shape below

Travel time:

Test #1 _____

Test #2 _____ RANK

Test #3 _____

Average _____

Transportation Systems

Alternate Laboratory Activity — Rubber-Band-Powered Boat

Name_____ Date _____

Period _____ Score _____

DESIGN BRIEF

Design a rubber-band powered boat that will carry a specified weight along a "canal." The boat that will carry the load the farthest with the "standard" rubber band will be considered the best design.

SPECIFICATIONS:

Maximum: Length _____   Specified load _____ ozs.

Width _____

Height _____

DESIGN

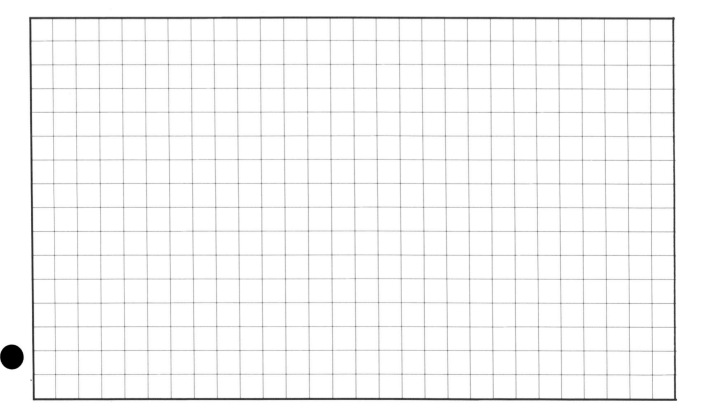

Chapter 11 — Study Questions

Text Pages: 219-245

Name _____

Date _____

Period _____ Score_____

1. Give two reasons why transportation is important to you.

    _____

    _____

    _____

    _____

2. Give three purposes for transportation pathways.

    _____

    _____

    _____

3. There are two basic types of land transportation _____-
    _____ - _____ and _____.

    3. _____
       _____

4. Trains are an efficient way to move people and goods because (select all appropriate answers):
    A. You meet courteous people on trains.
    B. Trains use cheaper energy than other modes.
    C. Tracks and vehicles are built to carry heavy loads.
    D. Trains do not get caught in traffic jams.

    4. _____

5. What is a hover craft and how does it operate?

    _____

    _____

    _____

6. _____ are types designed to transport loose materials or gases by suction or pressure.

    6. _____

7. _____ are long continuous belts or chains that move materials from one place to another. Sometimes they are used to move people too.

    7. _____

8. Name five types of vehicles used to move people or materials on water.

    8. _____
       _____
       _____
       _____
       _____

9. _____ include all airborne vehicles that travel within the earth's atmosphere.

9. _____

10. Explain how a helicopter flies.

_____

_____

_____

_____

11. Rockets carry their own _____ in addition to fuel because there is none in outer space.

10. _____

12. If you were to take an automobile, then an airplane, and then a ship as you travel, you would be using _____ transportation.

12. _____

13. __A__ and __B__ are structures that provide shelter for transportation activities.

13A. _____

B. _____

14. The following three events were most important to the historical development of transportation:
A. Law of gravity.
B. Development of the sail.
C. Invention and development of the wheel.
D. Invention and development of the heat engine.
E. The discovery and refining of petroleum.
F. Discovery of steam.

14. _____

15. What resources do people provide for transportation (list six)?

_____

_____

_____

_____

_____

16. Name the six components necessary to every vehicle.

16. _____

_____

_____

_____

_____

_____

17. Gas turbines, rocket, and gasoline engines are examples of _____ engines.

17. _____

_____

_____

18. List at least four ways that power can be transferred from one place to another.

_____

_____

_____

_____

19. A computer receives messages in the form of electrical signals   19A. _____

from ___A___. It sends orders to ___B___.   B. _____

chapter 12

Production/Biorelated Systems

Activity Sheet — Bread Making Report

Name_____ Date _____

Period _____ Score _____

In the space below, report your observations during the bread making activity and explain each observation.

OBSERVATIONS	EXPLANATIONS

Alternate Laboratory Activity — Cheese Making

Name _____ Date _____

Period _____ Score _____

ASSIGNMENT

Read the following description of cheese making. Then draw up a list of supplies and materials needed. Use the space provided. Next, in the space provided, develop a procedure for making cheese. (For additional information on cheese making, check the school Family and Consumer Sciences department, health food stores, or the school resource center.)

HOW CHEESE IS MADE

Cheese making is an ancient method of preserving milk, which, unless refrigerated, turns undrinkable in a matter of days. According to legend, it was first used by an Arab, who, in ancient days, set out on a journey. Since the journey was to be long, he took along milk in a container he had made from the stomach of a calf. When he attempted to drink the milk later, he found that it had turned to a solid. Finding it tasty, he consumed all of it and told his countrymen of his discovery.

This story illustrates the first steps in cheese making. Acid-producing bacteria must first make the milk more acid. Then the enzyme, rennin, which, even now, is obtained from the stomach of calves, coagulates the milk at a temperature of about 99 °F (37 °C).

In cheese making, the milk must first be slightly warmed. A starter is added to produce lactic acid. When the acidity increases, addition of rennin soon coagulates (curdles) the milk.

With curdling, certain liquids and water-soluble substances separate from the milk. This is called whey. Heat and the compressing of the curds can help in separating the two. Placing the curds in a cloth bag will make this task easier.

Uncured cheese is made from the curds of skim milk. It is only necessary to use a lactic acid starter, but rennin is often added for making cottage cheese with large curds. The curds are salted and cream added for creamed cottage cheese.

In the following chart, produce a list of utensils and supplies needed for making cottage cheese.

MATERIAL	QUANTITY

In the space following, give step-by-step instructions for preparing the cottage cheese.

PROCEDURE: _____

_____

_____

_____

_____

_____

_____

_____

_____

_____

_____

_____

_____

_____

_____

_____

_____

_____

Text Pages: 246-252

Name _____

Date _____

Period _____ Score _____

1. Resources of energy and material are never received from living matter. True or false?

    1. _____

2. Biotechnology is the _____ _____ of biological systems in processing materials.

    2. _____
       _____

3. Biotechnology uses living agents in an industrial process. True or false?

    3. _____

4. Which of the following would NOT be considered a bio-industry?
   A. Any business that deals with living things.
   B. Manufacture of mechanical hearts.
   C. Raising farm animals.
   D. Making cheese.
   E. A bakery.

    4. _____

5. What are biofuels?

_____

6. In a thermochemical process known as _____ heats biomass in the absence of oxygen.

    6. _____

7. What happens in the alcoholic fermentation process?

_____

_____

8. What recent discovery has produced a biotechnology "boom"?

_____

9. _____, in simple terms, is a set of plans for living organisms.

    9. _____

Impacts of Technology

Activity Sheet — Scrap/Waste Survey Chart

Name_____ Date _____

Period _____ Score _____

Survey your school for scrap or waste and place your findings on the form below. Additional space is provided on the reverse side.

WASTE AND SCRAP MATERIAL SURVEY*

ITEM	APPROXIMATE QUANTITY	LOCATION	TYPE
1.			
2.			
3.			
4.			
5.			
11.			
12.			
13.			
14.			
15.			
16.			
17.			
18.			
19.			
20.			

*Use with activity described on page 269 of the text.

Impacts of Technology

Alternate Laboratory Activity — Home Waste Report

Name _____ Date _____

Period _____ Score _____

Survey the garbage and waste generated in your home in one week. Make your report on the form below.

HOME GARBAGE AND WASTE REPORT			
Garbage (scraps of food, etc.) disposal method: [ ] Garbage disposal [ ] Garbage truck [ ] Burned [ ] Other (specify) _____ Trash (material other than food scraps) disposal method: [ ] Burned [ ] Removed by garbage truck [ ] Burned [ ] Other (specify) _____			
Analysis of waste (list items discarded)			
ITEM	**APPROXIMATE QUANTITY**	**RECYCLED**	**NOT RECYCLED**

CHAPTER 13 — Study Questions

Text Pages: 248-261

Name _____

Date _____

Period _____ Score_____

1. Consider the following activities and list one positive (good) effect and one negative (bad) effect each has on our lives.

   A. Building an expressway.

   Positive: _____

   _____

   Negative: _____

   _____

   B. Using an electric hair dryer.

   Positive: _____

   _____

   Negative: _____

   _____

   C. Taking students to school in a bus.

   Positive: _____

   _____

   Negative: _____

   _____

   D. Cutting lumber to build a bookcase.

   Positive: _____

   _____

   Negative: _____

   _____

2. Indicate briefly how your life would be affected if you did not have:

   A. Electricity: _____

   _____

   B. Packaged food: _____

   _____

   C. Refrigerator: _____

   _____

   D. Gasoline: _____

   _____

   E. Ready-made clothes: _____

   _____

F. Indoor plumbing: _____

_____

G. Garbage pickup: _____

_____

H. Telephone: _____

_____

I. Postal service: _____

_____

J. Radio: _____

_____

3. Name two things that we can do to save energy resources or scarce materials.

_____

_____

_____

4. What does the term, earth berming, mean?

_____

_____

_____

5. One of the greatest sources of air pollution is crop dusting. True or false?

5. _____

6. When is it good to delay the use of new technology?

_____

_____

_____

_____

_____

_____

_____

_____

_____

_____

_____

_____

_____

7. Write a brief report about a problem in your community. Tell how it affects you. If you know, report on any efforts the community is making to solve the problem or improve the situation.

_____

_____

_____

_____

_____

_____

_____

_____

_____

_____

_____

_____

_____

_____

_____

_____

_____

_____

_____

_____

_____

_____

_____

_____

_____

_____

_____

_____

_____

_____

_____

chapter **14**
Technology and the Future

Activity Sheet — Area Assignment

Name_____ Date _____

Period _____ Score _____

As a group, develop a list of living/working areas required on a space station. Then determine which floor of the space station each will occupy.

AREA ASSIGNMENT SHEET FOR SPACE STATION

Name of living or working area	Floor assignment

chapter 14
Technology and the Future

Alternate Activity — Future History

Name_____ Date _____

Period _____ Score _____

In the space below write a future history about what students in this course 50 years from now will experience. Talk about such things as what new technologies there will be, what textbooks might be like, how computers will be used, etc. (Refer to the discussion on future histories found on page 275 of your text.)

TECHNOLOGY EDUCATION IN THE YEAR 2050

_____
_____
_____
_____
_____
_____
_____
_____
_____
_____
_____
_____
_____
_____
_____
_____
_____
_____
_____
_____
_____
_____
_____
_____
_____
_____

Technology and the Future

Alternate Laboratory Activity — Ocean Wave Power Generator

Name_____ Date _____

Period _____ Score _____

In the future, our technology will require the development of different sources of energy. One source as yet untapped is the use of ocean waves. At present, our technology has been able to harness tides. However, there is considerable energy (kinetic) in wave motion.

Carefully study the design ideas that follow. Then read the problem (design brief) and the instructions for your activity on the next page.

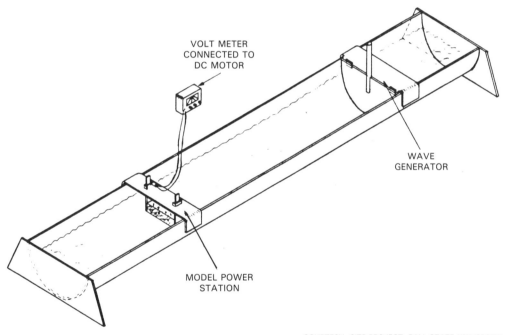

VOLT METER
CONNECTED TO
DC MOTOR

WAVE
GENERATOR

MODEL POWER
STATION

COURTESY, CITE PROJECT, BALL STATE UNIVERSITY

One way to study the effectiveness of full-scale designs is to test them on a scale model. The drawing above shows a model designed to produce and measure the electrical energy in the kinetic energy of ocean waves. One device produces the waves while another device converts the wave energy into electricity. A third device is needed to measure the amount of electrical current.

DESIGN BRIEF: Design a device that will produce waves whose energy can then be measured.
INSTRUCTIONS: Present your design ideas in the space provided. You may choose to work from a design suggested by your instructor. In the space provided, add your sketches, a Bill of Materials and a set of procedures.

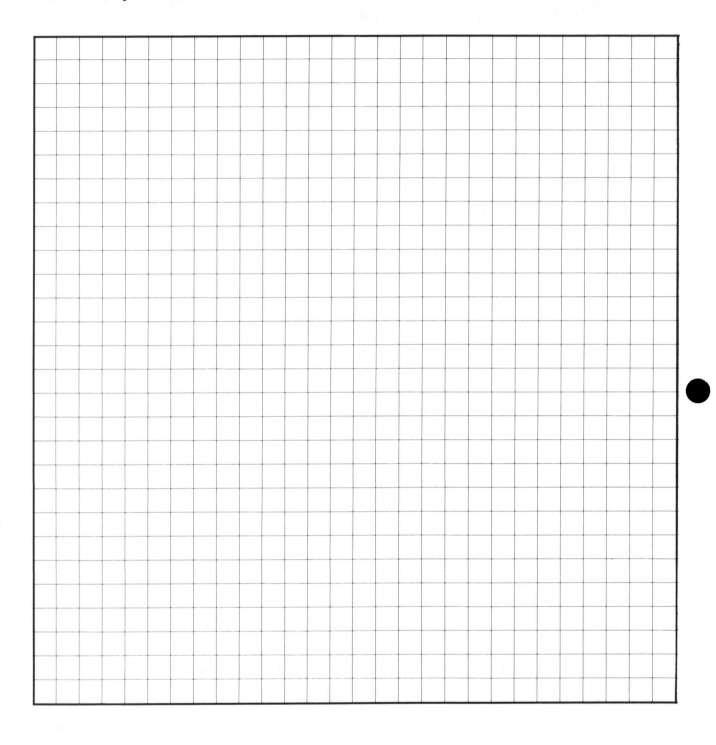

Technology and the Future

BILL OF MATERIALS

_____
_____
_____
_____
_____
_____
_____
_____
_____
_____
_____
_____
_____
_____
_____
_____
_____
_____

PROCEDURES

_____
_____
_____
_____
_____
_____
_____
_____
_____
_____
_____
_____
_____
_____
_____
_____
_____
_____

CHAPTER 14 — Study Questions

Text Pages: 270-280

Name _____

Date _____

Period _____ Score_____

1. What is superconductivity?

_____

_____

2. How do futurists predict what might happen?

_____

_____

3. A _____ projection deals with what may happen in the next couple years.

    3. _____

4. A long-range projection may look ahead as far as _____ years.

    4. _____

MATCHING TEST: Match the definitions with the terms.

5. Studies how events are moving in a certain field. Predicts where they will be in so many years.

    A. Modeling.

    5. _____

6. Compares events that are alike in some way. Assumes that what happened in one situation will also happen in the other.

    B. Networking and decisions.

    6. _____

7. Experts are asked to respond to a number of questions about the future.

    C. Collecting facts about trends.

    7. _____

8. A method of making things happen because a change is desirable.

    D. Use of analogies

    8. _____

9. Imagining yourself at some point in the future and then "looking" back at events that have not happened yet.

    E. Scenarios.

    9. _____

10. Making up a chart of the main events of a future happening. It allows planners to predict all the main events.

    F. Taking a survey.

    10. _____

11. Produce or describe a set of related events. Then give the consequences of these events.

    G. Developing a future history.

    11. _____